Signal contain some useful and important information and it is the function of some independent variable for example time or space .for example if we record the variation of room temperature with respect to time of any day and sketch a graph then this graph can be considered as a real world signal which contain some information and which is varying with respect to time (independent variable).Take an another example suppose there is a temperature source and as we move away from this temperature source temperature is decreasing whereas if we move towards this temperature source the temperature will increased. We can also sketch the variation in temperature with respect to space and this sketch can be considered as signal which is varying with respect to space (some independent variable).

Signal Processing : Signal processing means to process the signal (for example to perform some mathematical operations upon it) and extract some desired information from the signal.

Analog signal : If signal is a continuous function of time then it is called continuous time signal. Continuous signal are defined at every instant of time.

Discrete time signal: Signal which are discrete function of time or the signal which is defined only at discrete time instance is called discrete time signal.

Digital signal: Signal which has finite number of amplitude is called digital signal.

Binary Signal: Signal which has only two amplitude level is called binary signal.

DIGITAL SIGNAL PROCESSING: Digital signal processing is the subject of engineering in which we process the digital signal with the help of computer, perform some mathematical operation upon signal and extract desired information from the signal.

With the help of computer we can only process the digital signal. We cannot process analog signal or discrete time continuous signal with the help of computer because to store an analog signal or discrete time continuous signal in computer we do require an infinite storage capacity. Because analog signal can have infinite amplitudes define at every time (infinite time) where as discrete time continuous signal can have an infinite no of amplitude which are defined at finite (discrete) time.

So for digital signal processing first we need to convert an analog signal into digital signal. Any analog signal can be converted into digital signal with the help of following steps

[1] Sampling.
[2] Quantization.
[3] Encoding.

[1] Sampling: Sampling is the process of converting continuous time continuous amplitude signal into Discrete time continuous amplitude signal. Sampling is done with the help of sampler. Sampler is a high speed on off switch which can convert continuous time continuous amplitude signal into discrete time continuous amplitude .

Sampling is done at a frequency which is at least twice of maximum frequency component present in information signal.

$$f_s >= f_m \ldots\ldots\ldots(1)$$

Where $f_s =$ *Sampling Frequency*

$f_m =$ *Maximum frequency component present in information signal*

If sampling is done at a rate less than the twice of maximum frequency component present in information signal then signal information will be lost during the reconstruction of an original signal from its sampled version.

[2] Quantization: Quantization is the process of converting discrete time continuous amplitude signal into discrete time discrete amplitude signal. In the process of quantization we convert an infinite number of an amplitude into finite no of amplitude by approximating the sampled signal value .A quantization error is occurred In the process of quantization. The value of quantization error can be reduced by increasing the quantization level. But as the quantization level increases then we do require more no of bits to represents the signal sample i.e. the bandwidth will be increases.

So there is a Trade-Off between Quantization error and signal bandwidth.

[3] Encoding: In the process of encoding every quantized sample value is converted into the digital signal.

WHAT IS DIGITAL IMAGE PROCESSING

Digital Image processing deals with digital images performing mathematical operation upon it with the help of digital computer to extract desired information .

Digital Image processing is the branch of digital signal processing in which first we convert an image into the digital signal with the help of a digital camera and computer then store this image into a computer. Once image is stored into the computer then it is converted into the two dimensional matrix of numbers. Now we can perform any mathematical operation with these numbers and can modify or transform these numbers in to the desired way and can extract an information. This process of modifying or extracting the desired information from digital image is called digital image processing.

WHAT ARE THE DIFFERENT STEPS IN IMAGE PROCESSING:

The process of Digital Image signal processing consists of three steps

[1] IMAGE ACQUISITION

[2] PROCESSING THE IMAGE

[3] GET A PROCESSED IMAGE

[1] Image Acquisition: Image acquisition mean to capture a real world image with a digital camera or scanner and to store this image in a memory device. An image is consider to be consists of small size pixels where every pixels reflect a different light intensity. To convert any image into digital signal image is focused to the array of sensors with the help of lenses.

The resistance of these sensors are the function of light intensity falling upon it. So the value of resistance at every point on this sensor changes according to light intensity of an image. So the voltage at every point of sensor array will be developed according to the intensity of every pixel value of an image. Again these analog voltage is converted into digital with the help of Quantizer and encoder. To capture a real world image with digital camera and to store it into a memory device the two process is required

The Image is captured with the help of array of sensors. Every image can be conceder to be consists of Pixels. Pixel is the smallest element present in any image and each pixel can is represented by a number which is proportional to the intensity of that pixel element. The number by which the particular pixel is represented is depending upon that how many bits we are using for the representation of any number.

We cannot process an analog image with the help of computer because an analog image cannot be stored in a computer To convert an analog image into the digital image two process are required

[a] Sampling

[b] Quantization

Sampling:- Sampling means how many pixels we are using per square centimeter for capturing an image. Because we cannot use an infinite number of samples to represent an image otherwise the size of image becomes infinite and cannot be stored in a memory device.

With reference to the image processing sampling means that how many pixels are required to represent the image per square centimeter. Pixel is the smallest element of image. Suppose a picture of size 2 mega pixel means 2 million pixels are required to present the 1 square centimeter area of an image. 10 mega pixels means 10 million picture are required to represent the 1 square centimeter of the image. If the number of sample per centimeter square is more than image quality will be better. If the number of pixel is small per centimeter square then image quality will be degraded.

But if we choose more no of pixel then the size of image will be increased so it do require more memory space to store an image again to transfer or transmitted image over internet will be costly and it do require more time. So there is a tradeoff between these parameters. Again the number of sample depends upon our requirement and applications. For example if we want to count any object then we do require less no of sample per centimeter square but if we want to identify the detail of any object then more no of sample per square centimeter is required. For example suppose we are interested in counting only the number of car park then we do required less no of sample per square centimeter but if we are interested to read the number plate of any car then we do require more no of sample per square centimeter. As the number of sample per square centimeter increases then the

pixel size decreases and

Fig(1) Three pari images with different no of pixels per square centimeter.

As the number of sample per square centimeter decreases then Pixel size increases. The Pixels also sometimes known as DPI (Dots per inches) . Again if we use less number of sample or pixels to make a picture then we cannot Zoom our Picture more or we cannot take large print out of it.

1500*1200

750*600

300*375

150*188

75*94

38*47

24*19

10*12

Fig(2) Pari image with different no of samples(Pixels) per square centimeter

If we use more number of samples or Pixels to make our picture then we can easily zoom it or we can take larger print out of it.

Quantization: Quantization means how many bits we are using for representing the each sample value generally in a gray image we used 8 bits to represent a sample value.

Finally Image can be considered as two dimensional array where every value in that matrix shows the intensity of pixels. If the intensity of these pixels are denoted with the help of 8 bit then the values in image matrix can vary from 0 to 255 . if we shows the intensity of pixel with the help of 16 bits than these numbers will vary from 0 to 65535. Quantization means that how many bits are required to represent the each sample(Pixel) of an image. If we are using only 1 bit for the representation of Pixel element then we can have only two colors.1 for white or presence of light and 0 for absence of light.

Fig(3) Binary Image 2 color Image

Suppose we have 2 bits for the presentation of pixel intensity then we can have four colors or 4 shade or 4 gray levels .because 4 combination is possible with two binary bots

              00 for Black

              01

              10

              11 for white

Fig(4) Gary image with 4 gray levels

Similarly if we are using 3 bits for the presentation of pixel intensity then we can have eight possible combinations or eight possible colors (Shade)

000   For Black

001

010

011

100

101

110   and   111 For White

Fig(5) Image with 8 Gray Levels

Similarly with the combination of 4 bits we can have 16 different colors or Shade 000,0001,0010,0011,0100,1001,0110,0111,1000,1001,1010,1011, 1100,1101,1110,1111 and so on.

Generally the combination of 8 bits is used for the representation of Pixel values.

so we can have 256 possible combinations or 256 possible colors. "00000000" or decimal no "0" is used for the representation of **black** color. and "11111111" or decimal number "256" is used for the presentation of **White** color .

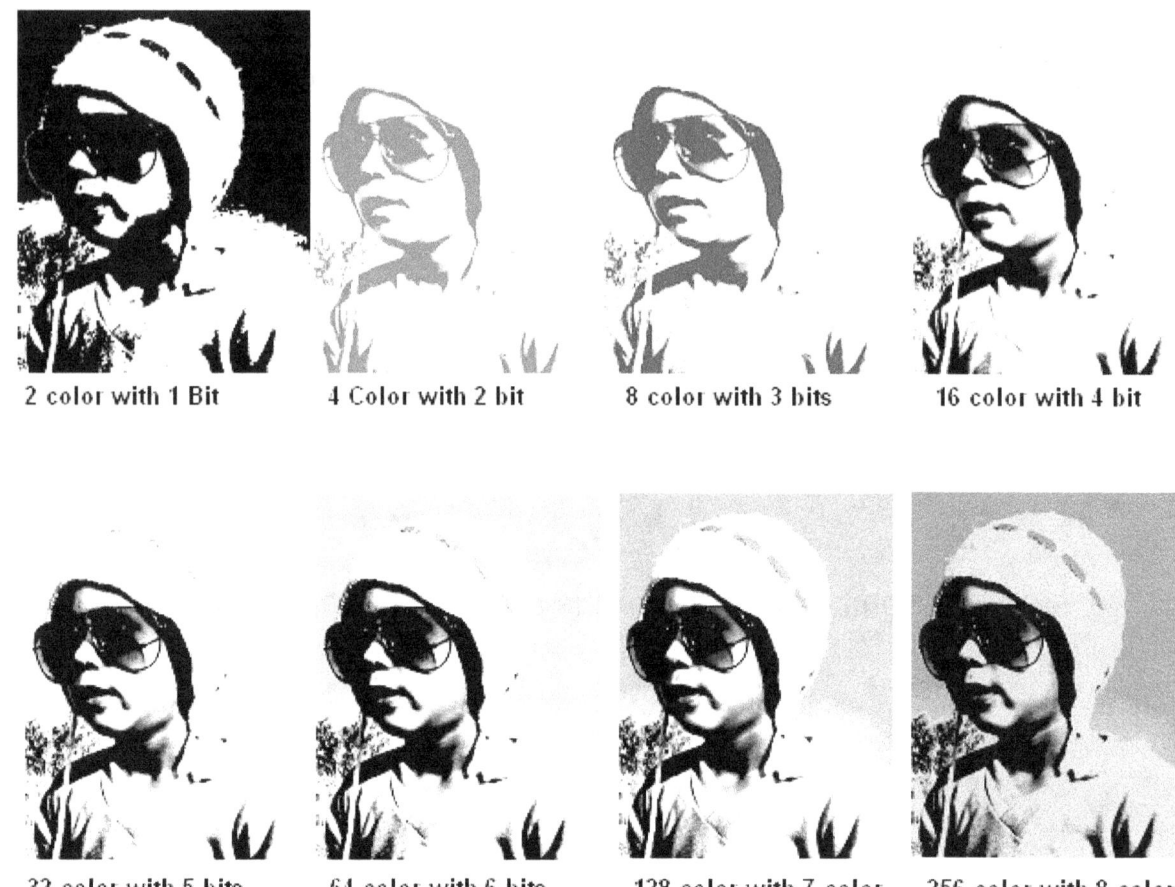

Fig (6) Pari Image with different number of shade or colors

[2] Processing the Image: Once Image converted into digital signal then we can consider this digital image as a two dimensional matrix of numbers and can perform any desired operation (Mathematical operation) upon it. So image Processing means to perform certain mathematical operation with the image matrix, or doing certain type of manipulation with image matrix and to either to extract the desired information from it or to get a better image from it.

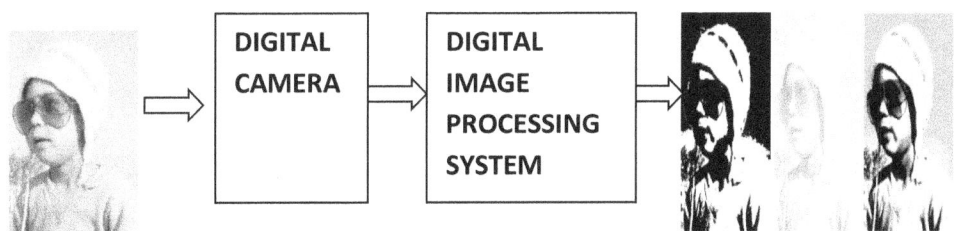

Fig(7) Digital image processing steps

[3] Get a processed image : Get a processed image means to get a better or desired image i.e. to get more brighten, more sharpen or better contrasted image.

We can also remove the noise and undesired frequency component from the image again we can detect the edge or segment the images into small objects.

OBJECT OF IMAGE PROCESSING

Image Enhancement

Image restoration

Image segmentation

Image Compression

Once we change the image into digital image and store it into the computer then image is converted into two dimensional matrix. Where numbers in two dimensional matrix shows the intensity of pixel elements of the image. Pixel is the smallest element of an image by these pixels an image is constructed.

In gray Image these numbers vary from 0 to 255(total 256 colors) when 8 bits are used to represent the pixel intensity and by changing the values of these numbers the quality of image can be varied in desired manner.

In 256 gray level image the number 0 is used for representing the black color or absence of light whereas the number 255 is used for representing the white color or presence of light.

We can increase the brightness of an image either by adding a constant value to Image matrix or by multiplying these image matrix by a constant number. and we can decrease the brightness of an image either by subtracting a constant number to these image matrix or by dividing this image matrix with a constant number.

In similar way we can also change the other properties of an image like contrast, sharpness etc.

Once we convert an image into the two dimensional array of numbers then we will be free to perform any mathematical operation over this image matrix and can get desired change in image

[1] Image Enhancement: Image enhancement means to improve the quality of an image. Which can be achieved by

[a] By Improving Sharpness of image

[b] By reducing the noise of an image

[c] De blurring the image

[d] Adjusting the brightness level and contrast of the image

[2] Image Restoration means to recover the original image with the help of damage image.

[3] Image segmentation means to divide the whole image into small objects and to do analysis with any small object of the whole picture or scene. Image segmentation is very helpful to extract some desired information from the image taken by satellite.

[4] Image compression: Image compression means to reduce the size of an image. Small size image is also helpful for the transmission over internet and to store in the memory of computer, camera or mobile etc. Image compression is done at the cost of image quality because to compress any image we need to remove certain image information.

Application of Image Processing

[1] Medical Science

[2]Face reorganization and fingerprints reorganization

[3] Pattern reorganization

[4]Industrial Automation

[5]Security Systems

Image size: With the help of an image matrix we can calculate the size of an image with the help of following formula

Image Size =no of rows in image matrix*no of column in image matrix * bits per pixel

Image Size = 1500*1200*24

Image Size = 43200000 bits

Image Size = 43200000/8

= 5400000 bytes

Again Image Size =5400000/1024

=5273.4375 Kilo byte

Again Image Size = 5273.4375 /1024

= 5.15 Mega Byte

TYPES OF DIGITAL IMAGES: There are three types of digital image

Binary Image

Gray Image

Color Image

[1] Binary image: Binary image can be considered as the two dimensional matrix where intensity of every pixel element can be represented with the help of single bit. In binary image there are only two color Black and White .Black color is represented by 0 bit and white color is represented by bit 1. So every Pixel of binary or black and white image is represented with the help of single bit.

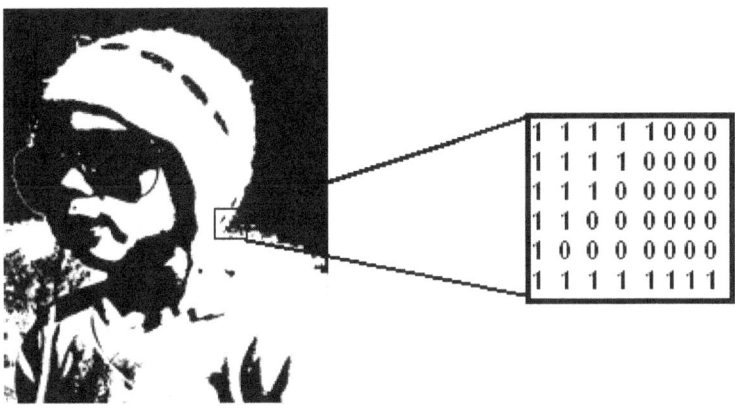

Fig(8) Pari Binary image

[2] Gray Image: Gray image can also be considered as the two dimensional matrix where intensity of every pixel element can be

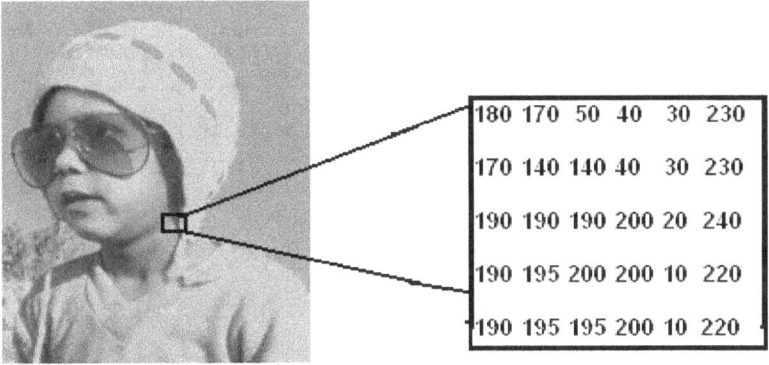

Fig(9)Pari gray style image

represented with the number which depend upon that how much bits we are using to represent a pixel. Gray image can have an infinite color it depend that how many bits we are using to represent a pixel element. With the help of 2 bits 4 colors or 4 shades are possible for every combination of bit 00,01,10 and 11 .

with the help of 3 bits 8 color or 8 shades are possible for every bit combination 000,001,010,011,100,101,110,111 .with the help of 4 bits 16 color or 16 shades are possible for every bit combination one different shade is possible for every combination of four bits and so on. With the help of the combination of 8 bits there are 255 colors or different shades are possible. 00000000 is used to represent Black color and 11111111 is used to represent white color.

[3] Color or RGB Image: In color image every pixel can be represented with the help of three numbers where each number corresponding to the intensity of RED, GREEN and BLUE color element.

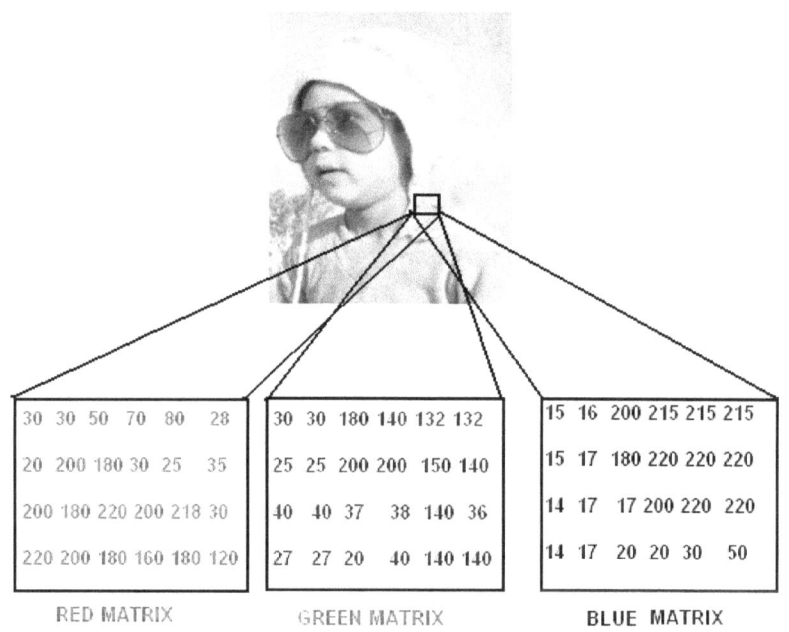

So a color image can be considered as the combination of three matrix each matrix is corresponding to one primary color. One matrix is for RED color, One Matrix is for GREEN and one matrix is for BLUE color.

POINT PROCESSING:- Point processing means to change an individual pixel value of an image. By point processing we can change the brightness and contrast of an image just by performing some simple mathematical operations with each individual pixel value of an image matrix.

CHANGING THE BRIGHTNESS OF AN IMAGE: Once we have converted the image into a matrix of numbers. Then we can perform any mathematical operation upon it. And by performing a mathematical operation we can alters these numbers (image) in a desired way. Simply by adding a constant value to the image matrix we can increase the brightness of image and by simply decreasing a constant value to the image matrix we can decrease its brightness.

Image

Image +50

Again by multiplying an image matrix with a constant no we can increase it brightness multiple times and by dividing any image matrix with a constant number we can decrease the image brightness number of times.

Image

Image-50

Image

Image*3

Image

Image/4

We can also do the same experiment with individual pixels of any image .

SIMPLE MATLAB EXAMPLES OF IMAGE PROCESSING

```
>> i=imread ('F:\Project\Image\pari.jpg');
>> figure,imshow(i)
   size(i)
   ans =
   1500      1200        3
```

>> imfinfo ('F:\Project\Image\pari.jpg')

ans = Filename: 'F:\Project\Image\pari.jpg'

FileModDate: '16-Sep-2015 20:24:32'

FileSize: 149665

Format: 'jpg'

FormatVersion: ''

Width: 1200

Height: 1500

BitDepth: 24

ColorType: 'truecolor'

FormatSignature: ''

NumberOfSamples: 3

CodingMethod: 'Huffman'

CodingProcess: 'Sequential'

Comment: {}

ImageDescription: '   '

Make: '   '

Model: '   '

Orientation: 1

Software: 'MediaTek Camera Application'

DateTime: '2013:11:30 08:41:34'

YCbCrPositioning: 'Co-sited'

DigitalCamera: [1x1 struct]

UnknownTags: [5x1 struct]


```
i(100,200,2)
ans = 217
>> i(100,200,1:3)
ans(:,:,1) =163
ans(:,:,2) =217
ans(:,:,3) =255
impixel(i,100,200)
ans = 171   221   255
```

MATLAB PROGRAM FOR CHANGING THE BRIGHTNESS OF AN IMAGE

```
>> i=imread ('F:\Project\Image\pari.jpg');
```

```
>> ig=rgb2gray(i);
```

```
>> figure,imshow(ig)
```

```
>>ig1=imadd(ig,128);
```

```
>> ig2=imsubtract(ig,128);
```

```
>> ig3=immultiply(ig,2);
```

```
>> ig4=imdivide(ig,2);
```

```
>> imshow(ig1),figure,imshow(ig2)
```

```
>> imshow(ig3),figure,imshow(ig4)
```

MATLAB PROGRAM FOR COMPLEMENT AN IMAGE

>> igc=imcomplement(ig);

>> imshow(igc)

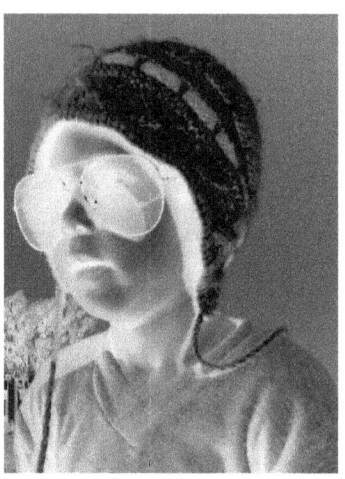

CHANGING CONTRAST OF AN IMAGE: Contrast means the difference between the intensity of pixel element of any image. If there is a more difference between the intensities between the darker and lighter pixels then we can say that the image has more contrast and if there is a less difference between the darker and lighter pixels than we can say that there is a less contrast.

HISTOGRAM: Histogram tells us the no of pixels belongs to any gray level. For example with 256 gray levels where 8 bits are used for representing the intensity of any pixel value. 0[00000000] number shows the absence of light or black color whereas 255 [11111111]number show the presence of light or white color. So as the numbers increases from 0 to 255 the brightness of pixels will be increased. So simply by observing the histogram of any picture we can know about the no of pixels belongs to any particular intensity level in that picture. If the histogram of any image is clustered at the left most side or 0 number side then the image will darker. Whereas if the histogram of any image is clustered at the right most side then the image will be brighter. Again if the histogram of any image is distributed within the interval of 0-255 it means that there are every pixel intensity present in our image. We can also increase the brightness level of any Image just by sliding the histogram of an image from left to right (from 0 to 255) and we can decrease the brightness level of any image just by sliding its histogram from right to left (from 255 to 0).

```
>> i=imread ('F:\Project\Image\pari.jpg');
>> ig=rgb2gray(i);
>>imshow(ig)
>> imhist(ig)
```

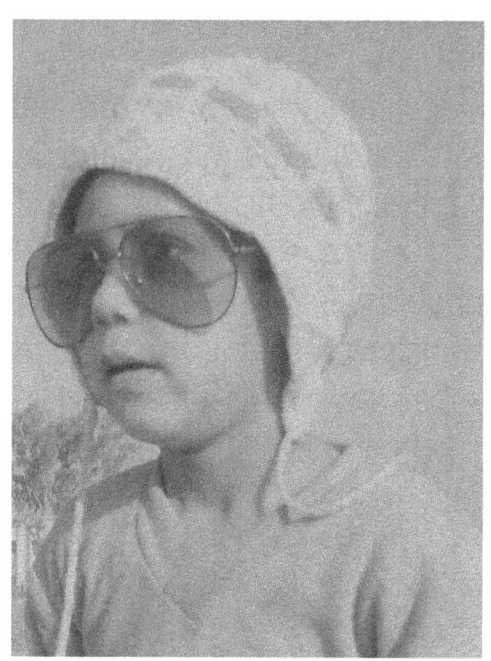

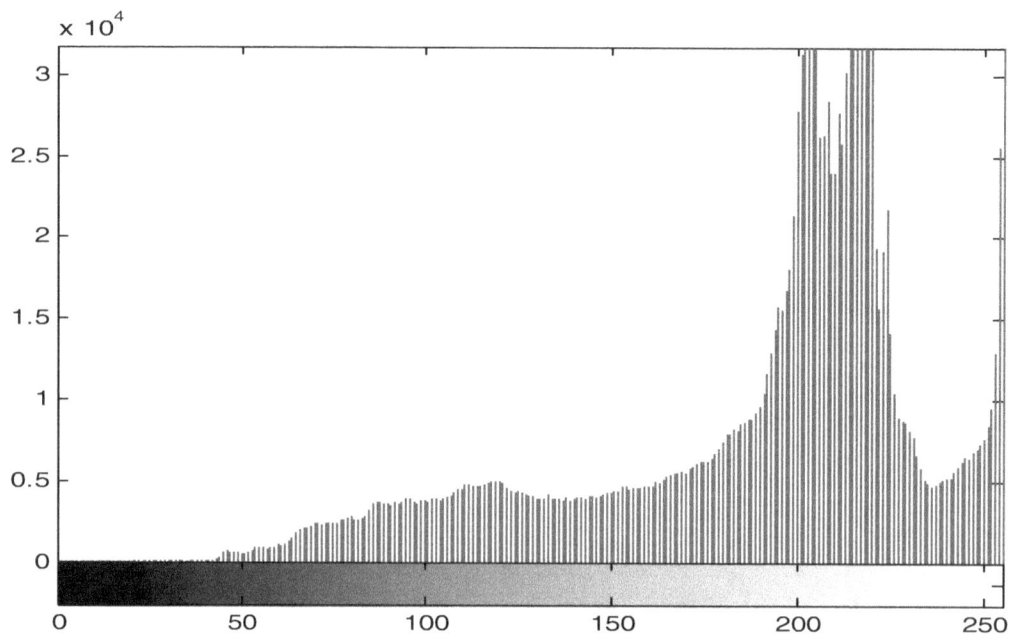

```
>> ige=histeq(ig);

>> imshow(ige),figure,imhist(ige),axis tight
```

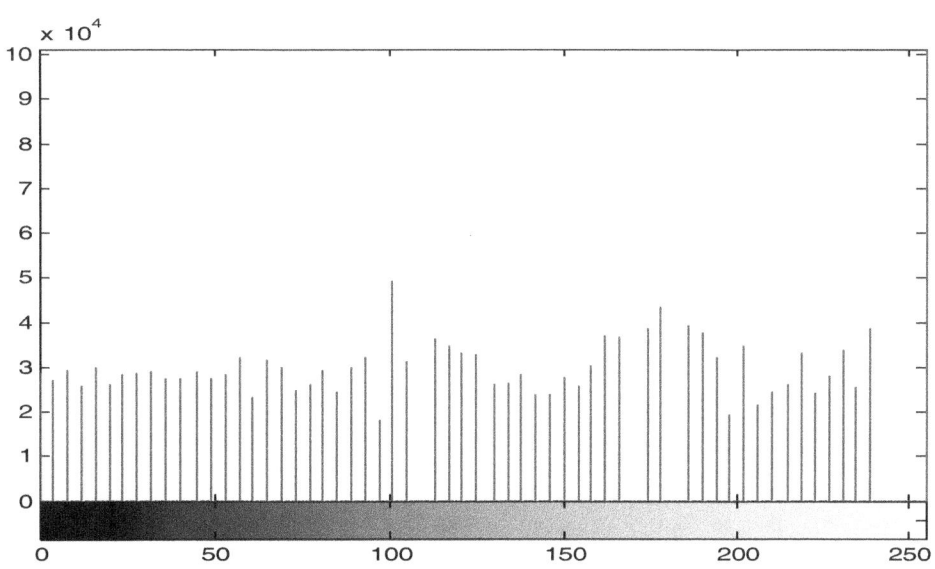

```
>>i=imread ('F:\Project\Image\pari.jpg');

>> ig=rgb2gray(i);

>>igad = imadjust(ig,[0.0;1.0 ],[0.5; .80]);

>> imshow(ig),figure,imshow(igad)

>> figure,imhist(ig)

>> figure,imhist(igad)
```

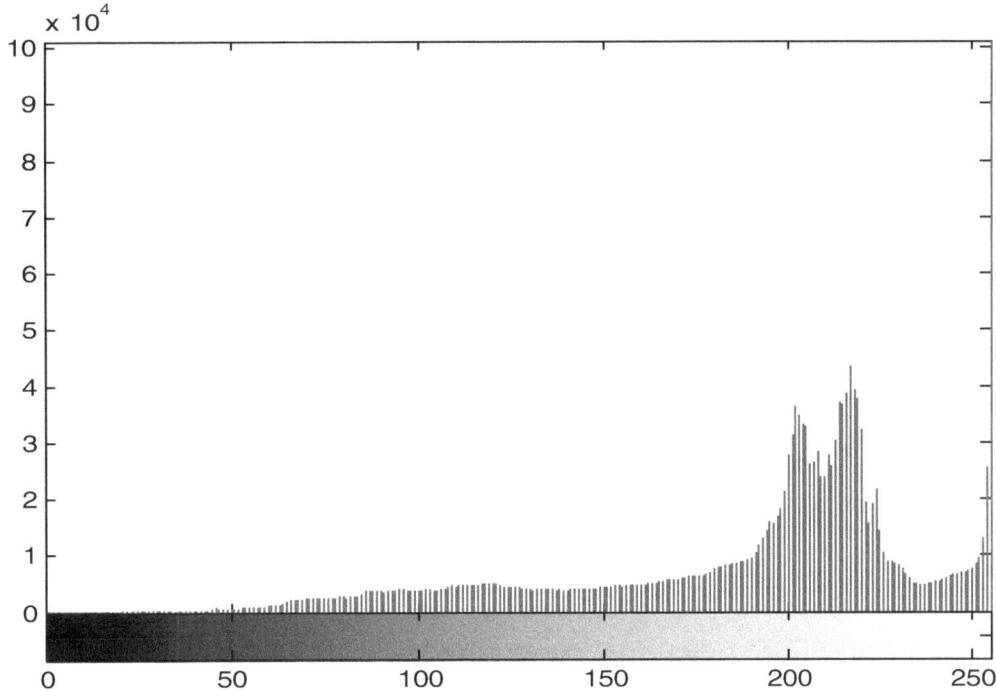

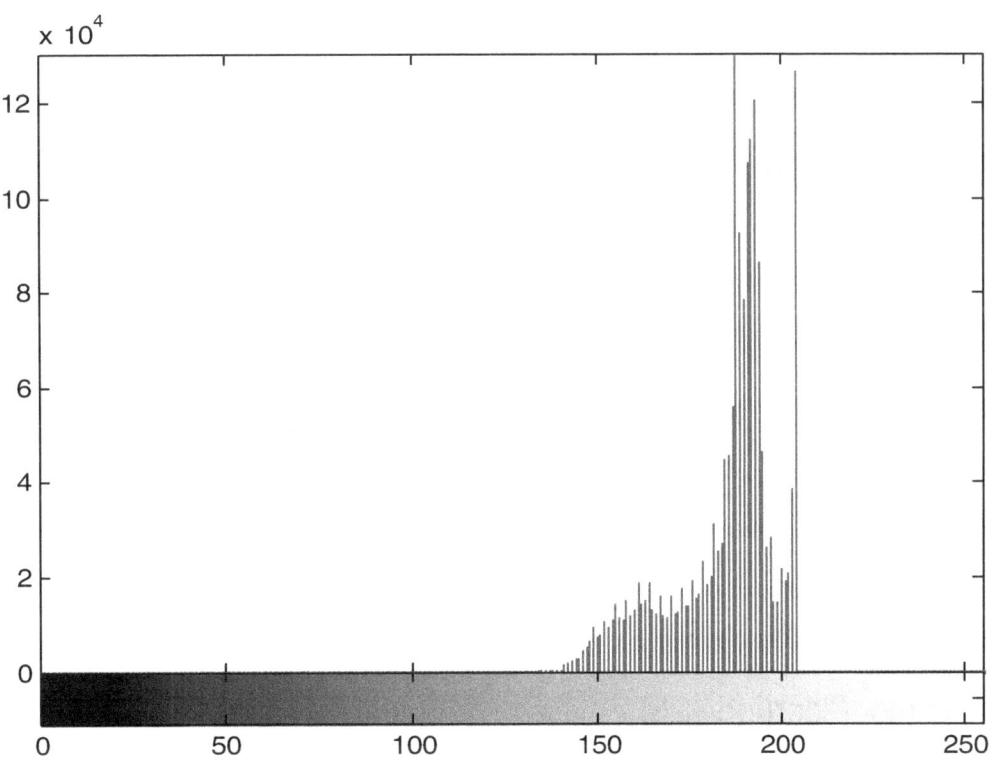

BIT PLANE: In a Gray Style image every pixel value can have a value between 0 and 255 and these pixel values can be represented with the help of 8 bit numbers. In these 8 bits every bit have some information about pixel and first bit (D0 position bit) contain least information where as eight bit(D7 position bit) contain most information about pixel.

So we can say that 8bit gray image can be break in to 8 plane and every plane consists of bit position from D0 to D7. An image consists of D0 bit position is known as least significant bit plane whereas an image consists of D7 bit position is known as most significant bit plane.

```
>> i=imread ('F:\Project\Image\pari.jpg');

>> ig=rgb2gray(i);

>> figure,imshow(i)

>> figure,imshow(ig)

>> igd=double(ig);

>> i0=mod(igd,2);

>> i1=mod(floor(igd/2),2);

>> i2=mod(floor(igd/4),2);

>> i3=mod(floor(igd/8),2);

>> i4=mod(floor(igd/16),2);

>> i5=mod(floor(igd/32),2);

>> i6=mod(floor(igd/64),2);

>> i7=mod(floor(igd/128),2);

>> figure,imshow(i0)
```

```
>> figure,imshow(i1)

>> figure,imshow(i2)

>> figure,imshow(i3)

>> figure,imshow(i5)

>> figure,imshow(i6)

>> figure,imshow(i7)
```

 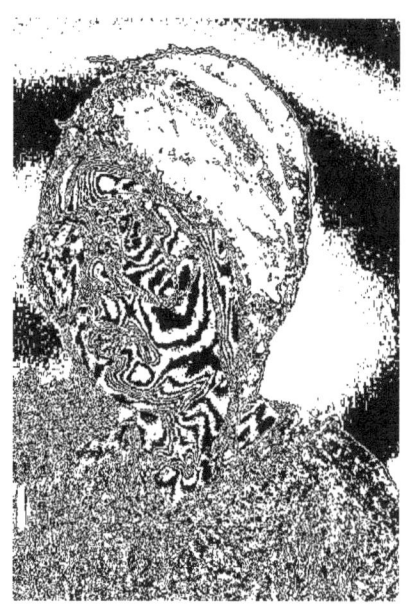

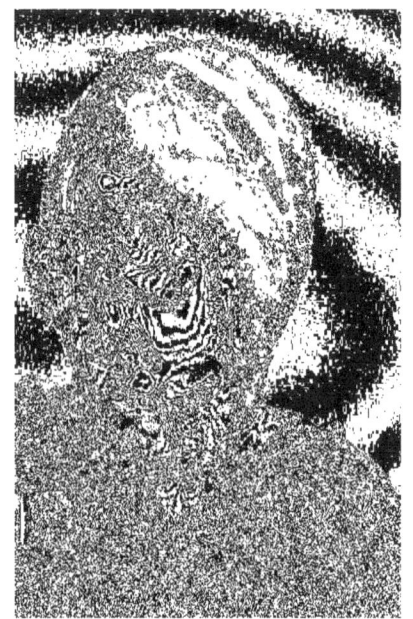 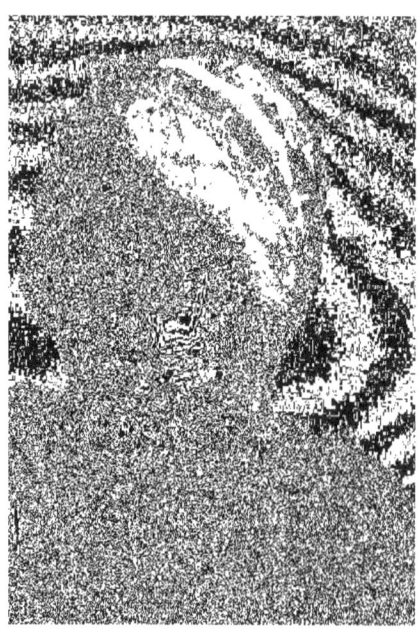

DIFFERENT FREQUENCY COMPONENTS' PRESENT IN IMAGE :

Different Frequency Components present in any image: As we know that when we convert any Image into digital image then it convert into two dimensional matrixes of numbers which values varies in between 0 to 255. These numbers shows the amplitude of Pixel at that point in the image. After analysis of these two dimensional matrix we can know that what different frequency components present in the picture and what are the actual location of these frequency components'

Low Frequency Coefficients: If in any small area the value of Pixels amplitudes are varying slowly then we can say that low frequency components' are present in that area of picture.

Small frequency components present in any image is also known as **approximate component** because it gives approximate information about the image these frequency components are very important and give major information about any image. The high frequency components are also known as **detail coefficient** because it contains detail information about any image.

High Frequency Coefficients: .If in any small area, the values of Pixel intensity changes very fast than we can say that high frequency component are present in that area.

Low pass filtering of an image: When image is passed through a low pass filter then its approximate coefficients are passed whereas its detailed coefficients are blocked. The simplest type of low pass filtering can be achieved by simply averaging the pixels values. There are lot of averaging low pass filters are possible for example 3*3 low pass filter, 5*5 low pass filter, 7*7 low pass filter etc.

There are two types of low pass filter.

LINEAR LOW PASS FILTER

NON LINEAR LOW PASS FILTER

Linear low pass filter: A linear low pass filter can be achieved by performing the linear mathematical operations on the pixel values of any picture. For example averaging low pass filter.

Averaging Filter: In averaging filter the pixel value of whole neighborhood(for example 3*3, 5*5 ,7*7 neighborhood) is replaced with the average value of pixels over that neighborhood

Some examples of linear low pass filters are

3*3 average low pass filter :

5*5 average low pass filter

7*7 average low pass filter

Non Linear low pass filter: Non linear low pass filter can be achieved by performing non linear mathematical function on the pixel value of any image. Some example of non linear low pass filters are

Median filters: In median filter the pixel values of neighborhood is replaced with

The median value of neighborhood. For finding the median value all the pixels are arranged in increasing order then all pixels are replaced with the middle order filter. If the number of pixel in any neighborhood is even then the pixel of whole neighborhood is replaced by the average value of two middle order pixels.

Rank order filters: In rank order filters all pixels of a neighborhood is arranged in increasing order and each pixel value assigned a rank then whole neighborhood pixel can be replaced with any rank order pixel in that neighborhood.

Minimum filters: In minimum filter the whole pixel neighborhood is replaced with the minimum value of pixel in that neighborhood.

Maximum filters: In maximum filter the whole pixel neighborhood is replaced with the maximum value of pixel in that neighborhood.

High pass filtering of an image: When image is passed through a high pass filter than its detail coefficient is passed whereas its approximate coefficients are blocked.

High pass filtering can be achieved by subtracting low pass filtering image from the original image.

ADDING AND REMOVING DIFFERENT TYPES OF NOISE FROM AN IMAGE

ADDING NOISE

[1] ADDING SALT & PEPEER NOISE USING MATLAB COMMEND

```
>> i=imread ('F:\Project\Image\pari.jpg');

>> ig=rgb2gray(i);

>> insp=imnoise(ig,'salt & pepper');

>> imshow(insp)

>> insp=imnoise(ig,'salt & pepper',0.2);

>> imshow(insp)
```

 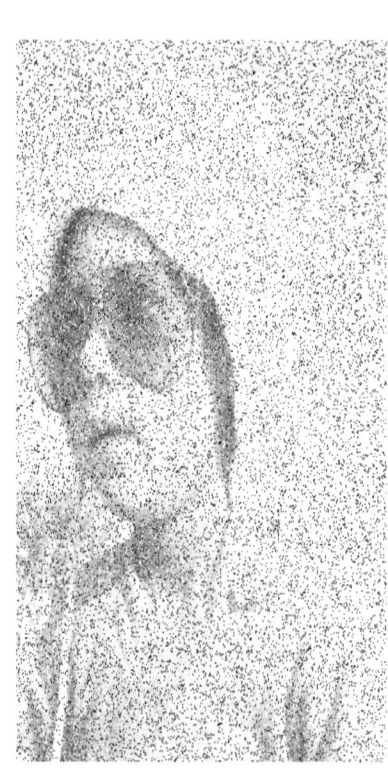

[2] ADDING GAUSSION NOISE USING MATLAB COMMEND

```
>> ingau=imnoise(ig,'GAUSSIAN');

>> imshow(ingau)
```

[3] ADDING SPECKLE NOISE USING MATLAB COMMEND

>> inspe=imnoise(ig,'speckle');

>> imshow(inspe)

>> inspe=imnoise(ig,'speckle',0.2);

>> imshow(inspe)

REMOVING NOISE:

REMOVING SALT & PEPEER NOISE:-We can remove salt and pepper noise using low pass filter. There are two types of low pass filter.

Linear low pass filter

Non linear low pass filter

Linear low pass filter: A linear low pass filter can be achieved by performing the linear mathematical operations on the pixel values of any picture. For example averaging low pass filter. Some examples of linear low pass filters are 3*3 average low pass filter

5*5 average low pass filter

7*7 average low pass filter

Non Linear low pass filter: Non linear low pass filter can be achieved by performing non linear mathematical function on the pixel value of any image. Some example of non linear low pass filters are

Median filters

Rank order filters

Minimum filters

Maximum filters

Matlab example of removing salt & pepper noise using median filter :

```
>> i=imread ('F:\Project\Image\pari.jpg');

>> ig=rgb2gray(i);

>> igsp=imnoise(ig,'salt & pepper');

>> igmed=medfilt2(igsp);

>> figure, imshow(igsp), figure, imshow(igmed);
```

MATLAB EXAMPLE OF REMOVING GAUSIAN NOISE USING ADAPTIVE FILTERS

```
>> iadp=wiener2(iggau);
>> imshow(iadp)
>> iadp1=wiener2(iggau,[5,5]);
>> imshow(iadp1)
>> iadp2=wiener2(iggau,[7,7]);
>> imshow(iadp2)
>> iadp3=wiener2(iggau,[9,9]);
>> imshow(iadp3)
```

IMAGE SEGMENTATION: - Image segmentation means partition of an image into small separate objects or to divide any big image into small component parts. Image segmentation can be studied under two subtopics

[1] Image Thresholding

[2] Edge detection

IMAGE THRESHOLDING:-With the help of thresholding we can convert gray style image into black and white image

By choosing a threshold value T a gray pixel value can be converted either into black pixel or white pixel.

If any pixel value is greater than threshold value then it is converted into white otherwise it is converted into black

There are two types of thresholding

SINGLE THRESHOLDING METHOD : By using single thresholding method we can convert any gray style image into binary image. By choosing a thresholding level we can convert each gray style value into either 0 or 1. If pixel value

is greater than threshold than it will be considered as 1.If pixel value is less than threshold than it will be considered as 0

Example : As we know that 8 bit gray scale image can have any pixel value from 0 to 255.

If we choose threshold value 127 Then any pixel value < 127 will be considered as 0 and any pixel value > 127

will be considered as 1

Some matlab examples of thresholding

Single threshold:

i=imread ('F:\Project\Image\pari.jpg');

ig=rgb2gray(i);

imshow(ig),figure,imshow(ig>100)

imshow(ig),figure,imshow(ig>200)

imshow(ig),figure,imshow(ig>150)

imshow(ig),figure,imshow(ig>170)

```
>> i=imread ('F:\Project\Image\pari.jpg');

>> ig=rgb2gray(i);

>> im2bw(ig,0.43)
```

```
>>i=imread ('F:\Project\Image\pari.jpg');

>>ig=rgb2gray(i);

>> im2bw(ig,0.39)
```

DOUBLE THRESHOLDING METHOD: In double thresholding we choose two threshold values T1 and T2 . We can implement double threshold in any gray style image. In double threshold method If the value of any pixel lies in between T1 and T2 than it will be considered as 1 (white)otherwise it will be considered as 0 (Black).

For example as we know in any gray style image a pixel value can vary from 0 to 255. Pixel value 0 is for black color and Pixel value 255 for white suppose we choose two threshold value 100 and 200. Than any pixel value which exists between 100 and 200 will be converted to 255 (white) and any pixel value which is less than 100 and greater than 200 will be considered as 0 (Black).

MATLAB EXAMPLE OF DOUBLE THRESHOLD METHOD USING MATLAB

i=imread ('F:\Project\Image\pari.jpg');

ig=rgb2gray(i);

imshow(ig),figure,imshow(ig>170 & ig<230)

ADAPTIVE THRESHOLD: Sometimes it is not possible for any image that we choose single threshold or double threshold value in such cases we use adaptive threshold. for example if brightness level of any image is changing continuously with respect to the space than by using single threshold value we cannot convert image into black and white image in such cases we use adaptive threshold where the threshold value changes with respect to the space

APPLICATION OF THRESHOLDING:

[1] In certain application we are interested in number, size or shape of an object then we can remove all the unnecessary detail from an image just by turning each gray level either into 0 (black) or 1 (white).

For example suppose we want to count number of cars in parking and we are not interested in color or shape of a car. Than we can remove an unnecessary detail just by converting the each pixel value into black and white.

[2] Sometimes because of similarity in gray level of pixel values detail of an image is hidden so threshold is used for revealing the hidden detail of an image .

EDGE DETECTION :- Edge detection is very important, essential and useful topic of image processing by detecting the edge of an image we can know about the shape and size of an image again by detecting the edge we can divide any image into small individual objects so we can count the number of objects in any image.

Edge detection is also helpful in face reorganization and pattern reorganization applications.

EDGE:- At the edge of any image the pixel value changes suddenly . Just by observing the image matrix we can know that where the edge of an image exists

120	121	120	10	130	125	110
122	122	120	10	137	124	119
115	118	121	09	135	123	118
116	117	120	11	134	130	117
118	121	120	12	135	131	118

For example in the above image matrix as we move from 3rd Columns to 4th columns the value of pixel changes suddenly so we can easily understand the location of edges.

So there are two types of edges

[a] Sudden edges   (Step): If the pixel value changes suddenly it is known as sudden edge.

[b] Gradual edges (Ramp): If the pixel value changes gradually then it is known as gradual edge

Mathematical operation for finding an edges: By using derivative operation we can find the edges in any image matrix.

[1] First derivative: Example of first derivative

f(x+1)-f(x)/1

Or   f(x)-f(x-1)/1

Or   f(x-1)-f(x-2)/1

[2] Second derivative

f(x+1)-f(x-1)/2

f(x)-f(x-2)/2

f(x+2)-f(x)/2

are some example of second derivative

Just by taking the derivative of an image matrix we can know that where the pixel values changes abruptly and where

The edge of an image exists.

EDGE DETECTION FILTERS:

Using first derivative

[1] Prewitt filter

[2] Robert cross gradient filters

[3] Sobel filters

Using Second derivative methods

[1] Laplacian

[2] Zero crossing

MATLAB EXAMPLE FOR EDGE DETECTION:

USING PREWITT FILTER

>>i=imread ('F:\Project\Image\pari.jpg');

>>ig=rgb2gray(i);

>> igepre=edge(ig,'prewitt');

```
>> imshow(igepre)
```

```
>> igerob=edge(ig,'roberts');
>> imshow(igerob)
```

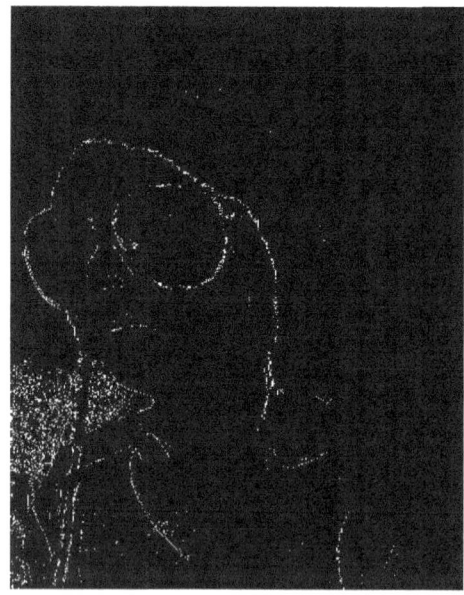

```
>>i=imread ('F:\Project\Image\pari.jpg');
>>ig=rgb2gray(i);
```

```
>> igesob=edge(ig,'sobel');

>> imshow(igesob)
```

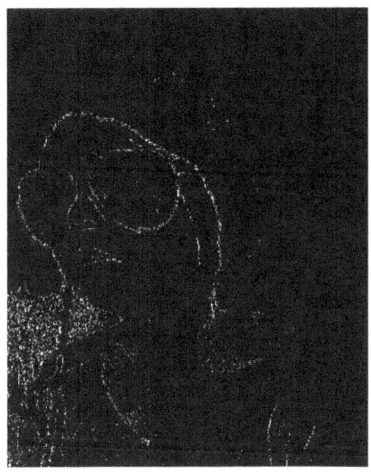

```
>> i=imread ('F:\Project\Image\pari.jpg');

>> ig=rgb2gray(i);

>> iclog=edge(ig,'log');

>> imshow(iclog);
```

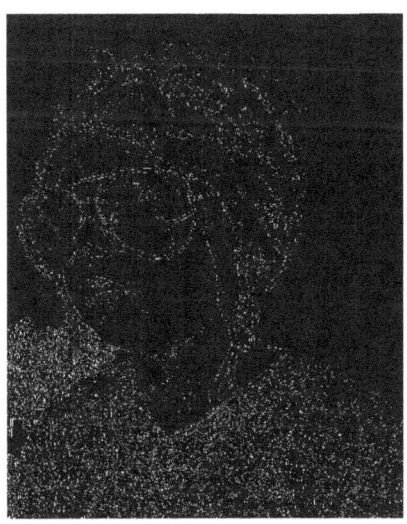

THANK YOU